SELECTED POEMS

ALSO BY HUGO WILLIAMS

Writing Home (OUP 1985)

SELECTED POEMS

Hugo Williams

Oxford New York
OXFORD UNIVERSITY PRESS
1989

Oxford University Press, Walton Street, Oxford OX2 6DP
Oxford New York Toronto
Delhi Bombay Calcutta Madras Karachi
Petaling Jaya Singapore Hong Kong Tokyo
Nairobi Dar es Salaam Cape Town
Melbourne Auckland
and associated companies in
Berlin Ibadan

Oxford is a trade mark of Oxford University Press

This selection first published 1989

British Library Cataloguing in Publication Data
Williams, Hugo, 1942–
Selected poems.—(Oxford poets).
I. Title
821'.914
ISBN 0–19–282563–1

Library of Congress Cataloging in Publication Data
Williams, Hugo, 1942–
[Poems. Selections]
Selected poems/Hugo Williams.
p. cm.
I. Title.
821'.914—dc19 PR6073.I432A6 1989 88–21898
ISBN 0–19–282563–1 (pbk.)

Typeset by Wyvern Typesetting Ltd.
Printed in Great Britain by
J. W. Arrowsmith Ltd., Bristol

To Hermine

Contents

From *Symptoms of Loss* (1965)

From *Sugar Daddy* (1970)

From *Some Sweet Day* (1975)

From *Love-Life* (1979)

From *Writing Home* (1985)

Symptoms of Loss

Don't Look Down

Don't look down. Once
You look down you own
The fall in your heart,
You rock your stance
On the stone
And hear your ending start.

It climbs up to you
Out of a deep pit
In sentences
Half followed through
And sighs which twist it
Till it wrenches

At your hold, or shows
You as a clown
Whose imitation rage
Will draw from rows
Of seats some laughter, thrown
Like nuts into a cage.

While high in the big top
A white-clad flyer springs
And it is he you were
Who made men stop
And wonder at your wings
So famous in that air,

Until you looked down
And saw your future there
In the dust and light
And suddenly were thrown
Into that pit where
Now you trot each night.

The Net

Accidents will happen. Swiss
Watches smash themselves
On pavements, rockets miss

The moon and noes and yesses
Get confused in tests,
Answers become guesses,

Till the ministry of friends
Is not so sure it's all
So simple, that it ends

In failure. They must let
You know they want to help,
Erect a kind of safety net.

And then the trouble really
Starts: the mesh too gross
For explanations, nearly

Always misses abstract nouns.
Arms fast, legs caught, your words
Go crashing down among the clowns.

Still Hot from Filing

The key is still hot from filing, silver
Against skin, like an arrowhead
And sharp. I close my hand over

Its bright surface and feel the fresh-
Cut notches dig into my palm,
Sterile within the wards of flesh,

But purposeful. It leaves its mark
And I become addicted to
The roughness, resolute as bark,

Or wire. I test the hardness of my nail
Against the barbed edge and recognize
A need to understand in Braille

My own antithesis, or that to which
I kneel. Then knowing what it is,
I turn and, keylike, feel an itch

To press my sharpened faculties once more
Upon more malleable stuff,
To watch my influence unlock a door.

The Pick-up

The hardest part is conversation. That is
Providing they both survive
The first shock of assault, the initial
Jolt of finding in a motive

Something more than interest in the bus-
Routes. (A moment when their manner
Falters into a new role.) Both know they walk
Tightropes like duellists and to gore

The enemy is to fall on one's sword.
Both guess at something more, the need
To escape. But the rules allow them
No more help. Now one must take the lead

And the hardest part begins. The same part
Played by the poet, all his senses alive
To the schemes in his head,
The twicking of his thumbs, who yet cannot give

Through a certain reluctance to lie,
To commit himself. A virgin, his first
Words on the page will comprise
Their anxious overtures, carefully nursed

Into something like sanity by similar
Desires: escape and the poem laid neat
In its place. Words having served their course,
The heat goes back to its simple beat.

The Actor

Sent by his agency to this bright
Box, all his lines straight, knew
Where to die, but was not quite

Sure who he was, or where he'd got
His hat and whether those breathless eyes
Were just his own, each pupil a dot

Of sanity or irony. In the cupboard
His uniforms were headless ghosts:
First, second, third, all safely stored,

Progressing through the acts alone.

Each of them a little more ripped
(Signifying war) than the one before.
The first being left almost undipped

In blood. Tools of his trade, they stand
Representing the passage of time: past,
Present and future, ranged close at hand

Upon their hooks, while not far away
Our soldier sits, uncertain of which one
To wear and whether it's all in play.

Right Moments

The right moment doesn't come
Ticking inevitably round like teatime.

It's not a dental assistant saying
'Mr Williams were you waiting

For the right moment of the day?
Would you very kindly step this way?'

Right moments haven't got time
To play the waiting game.

They picnic in the rain.
They would do the same again.

And not inside the head.
And not underneath the bed.

But suddenly, without consideration
For the opinion of the population,

Their exacting, clenched desire
Shall be let out into the air.

For right moments don't happen
If wrong ones are never taken.

Right moments are pram covers
For those who have no others.

Driving on the A30

I

Now I hold exact location
Of my life within a single action.

This, the matched coherence of my route
And eye, the pressure of my foot.

It puts me there beyond the waste
Of nothing ventured, nothing faced.

And keeps me bold and fisted
Like a hawk which ranged and twisted

As we passed, then stooped to wrest
Its victim from a field. The best

In all of us is what we do
When forced to find an answer to

The will which thrusts us forth
In time of urgency, not aftermath.

II

Now it is dark. I only see
The outline of your head is near to me

And pale as headlamps briefly
Press into the gloom then swiftly

Pass. I try to see if you are smiling
And forget to dip. Immediately unsmiling

Drivers flash their hate at me. I glare
Mine back. It makes a fair

Analogy of what we offer in return
For smiles and scowls and how we earn

Our praise. Now I can see you sleeping
Next to me. A yellow light is falling

On your hands. I try to memorize
The picture as we move, but cannot close your eyes.

Symptoms of Loss

It gives off a smell of burning
At first. No definite pointer.
Just a nudge from the sixth sense
And then nothing at all, the way
Instinct yields temporarily to
Reason. The lull before the fray.

The next stage is not so peaceful.
It consists of a furious hunt
In some remote quarter where it is safe
To search without hope: our fears allow
Us there. You'll see we've
Hidden from the truth till now.

This too is the way we react,
How nature postpones the fact for us
Until we are more obedient to loss,
Almost passive. The right places
Then are left until the end,
When the pulse no longer races.

Delphi

Light is present in this valley
  As in no other. It is made of green
And black and comes from the sea.

  There is snow on the cliff face
And in the air, but you can see every leaf
  On the olive trees at the base

Of the mountain and a splinter
  Of rock like a bone on the opposite
Precipice. Light is empty in winter

  And throws back images of distant
Animals and birds which turn and look
  At you from miles away with vacant

Eyes and their wings hardly moving.
  It is part of a stillness
Which is buried in the hillside. Something

  Almost submarine about a silence
Of stones and the ghosts of temples
  And stadiums rising from their immanence.

Women of the Nile

The women of the Nile wear gold
  Under the black. Bright crescent moons,
Three in each ear and the fold

  Of a shawl may hide large axe-
Like diadems, braided gilt. They seek
  To catch about them what the desert lacks

In brilliance of the shining perfect
  Surfaces they shield from the sun's
Cheap metal and the disrespect

  Of men. They cloak their own beauty
In a similar doubt, as if
  Through sight a theft might be

Accomplished, or a kiss. Jewels they must
  Take safely to the grave,
Where ornaments uphold the dust,

  But lips and cheeks they bring
To a husband's bed, who knows
  That a glance is valued like a ring.

Crossing a Desert

This truck puts an end to dreams,
How we arrive in great cities
Simply by wishing on their names:
Bassora, Isfahan, casting a spell
On our senses, a coppery bell
Of their syllables. Or lie awake
Nostalgic for what might have been,
Reluctant to invade the wasteland
When all the leave-taking is over,
The image out of hand.
Up here the din is unmistakeable: life
Aimed out of the silent shadows
Along its own path, a track into the last
Rays of sun, the meadows of dust.

Beginning to Go

I watch your complicated face
In a three-sided looking-glass,

Intent on a radio serial
As you pile the subtle

Darkness of your hair, each morning
Higher on your head. Last evening,

After two bottles of beer
We almost spoke: your sister

Manufactures silk in Bangkok.
She gets about £2 a week

And it's no bloody good. Your own
Work here is harder to explain.

We laugh at almost the same thing,
Uncertain whom the joke is on.

The rift is here already, though we laugh
At it. And though I laugh,

I feel the dried-up sadness
Of it, like age coming into my face.

Aborigine Sketches

The black men hang their shadows
On ropes underneath the towers.

Their hats slip forward over their eyes
Like the hats of lynched men.

They have been left standing these
Few people, like the dead gum trees,

Grotesquely upright still,
But slowly whitening.

*

The Mission is embalmed in charity.
The dreams of dead
Misguided German Christians lie an inch
Under the dusty sand that will never
Be sown or broken with laughter.

All day the families of matchstick children
Shift like hour hands round eucalypts.
Hazed in flies, a bleary wolfhound
Shambles across the courtyard—ratbag
Mascot of some disgraced regiment in exile.

*

They hold out to us
Discredited skilled hands they have lost faith in.
We prune them back like jungle for the public good.

This silent Reserve, their country in Arnhem Land,
Is a lopped hand on each of them. Their hands
Are disappearing into the desert and the dreamtime.

*

.e is only beautiful
.n the manner of his country.
.He was burnt by the same enemy,
Was at the same treaty.

And now he lives on a concession
Which shifts with the season,
That he must follow it
As a jackal follows his lion,

Licking at gnawed bones
Till he himself is one,
Hollow and dry as an old tree,
Full of strange, delicate energy.

He can walk a whole month
Into the desert in the dreamtime,
Can scent water on wind
And make rabbits jump into his hand.

He can hit a snake with a stone,
Can play a long, sad note
And burn stories on bark,
Beautiful in the manner of his country.

*

They came to us like lepers
To be cured of nightmares

And we woke them up
And showed them their sores

And hung them like flypapers
In museums called Missions

And said 'We are barely
Keeping them alive', as if

We regretted the way they sometimes
Stirred up their own dust

With a little rum
Bought for them by a tourist.

*

A child sulks upon the lawn,
Exhausted by the lawn's piety.

She wants to take off her dress
And sink into the earth.

For the pink roses have twined
Braceleted arms round her neck

And roots round her body,
Sapping her strength.

The Stage is Unlit

come back late at night from your room.
walk up that steep road
Which looks from the sea like a vertical plank.

My room is cold. The door bangs.
Neither of our rooms has a door which closes.
Mine keeps blowing open. Yours won't shut.

We both have balconies. Yours looks on to a yard,
Mine over roofs to the sea.

have a better room than yours, but no electricity.
use old kerosene lamps which smoke and have to be
  trimmed.

In your room we don't use the light,
For you have a son, not to be woken,
In his blue cot hung with rugs you've made him.

The Coalman

The coalman is in the street
And his street-cry is in the houses

Where people are thinking about coal
And their need of it. They think about

The coalman and how his head
Bears all the weight of the coal,

His eyes descending with it into the area
While a lorry creeps along the street.

And they look forward to drawing their curtains
And examining the Amusement Guide

And forgetting the world outside
With all its tiresome two-note warnings

And the coalman going past their doors
With his knowledge of their wretched fires.

The Butcher

The butcher carves veal for two.
The cloudy, frail slices fall over his knife.

His face is hurt by the parting sinews
And he looks up with relief, laying it on the scales.

He is a rosy young man with white eyelashes
Like a bullock. He always serves me now.

I think he knows about my life. How we prefer
To eat in when it's cold. How someone

With a foreign accent can only cook veal.
He writes the price on the grease-proof packet

And hands it to me courteously. His smile
Is the official seal on my marriage.

Sugar Daddy

Builders

A cage flies up through scaffolding
Like a rocket through time.

Thirty-six floors,
The numbers in white on the windows.

High on the roof I see men clearly
In their yellow helmets, talking.

One of them laughs at something on the river.
A negro turns out his pockets.

Slowly a crane goes by,
Dragging a name through sky.

Brendon Street

I watch the back of the casino: precast walls
Stained black already where
Last year a terrace stood like ours.

In the loading area: ashcans, sports cars,
Scaffolding joints, some mauvish masonry blocks,
A detective paring his nails.

A coiled hose spurts little floods
Of water on the pavement. The brass nozzle
Seems to move away backwards of its own accord.

A van arrives, reversing in a wide curve
To the lift gates where some small gilt chairs
With buttoned seats are waiting.

At seven, the croupiers bristle forth
With cigarettes, handling lighters.
These are the lords of Brendon Street. Their shoulders

Barge against the evening like a ball and chain.
They shoot white cuffs
And kick bright patent leathers this way and that

Among the empties and cats.
A concierge in trousers bows to them
And drags her poodle down the road. A girl I've seen

Looks at my window, but I can't be sure.
I could not move to follow her if I tried.
I stare out through my tent-flaps like a squaw.

In a Café

Sometimes the owner's mother
Comes out from the back like a stranger.

She can still take money for things
But keeps it clenched in her fist and has to be helped.

She fills her cup from the canteen,
Lights a cigarette, inhales, you can see

From the way she draws back her lips
That revival withers her.

I almost see it staining her skin
Like vinegar through newsprint on the floor.

Motorbike

The saddle is frozen solid.
The chronically wet rubber sponge
Inside the leopardskin cover
Crunches like shingle.

I hold my cuff
And wipe off the surface rain,
Lean over and flood the carburettor,
Jump on the start again.

A sneeze.
A little plume of steam.
The old tubes cough up a bit of phlegm
Then fade.

I have chronic catarrh, a raw ankle,
Pinkeye, blackheads and foul hair.
I have a humiliating sheepskin coat
And I lust strangely after a new alternator.

Sure

Walking upstairs after breakfast
I looked round to see if you were following
And caught sight of you
Turning the corner with a tray
As I closed the bathroom door.

The Couple Upstairs

Shoes instead of slippers down the stairs,
She ran out with her clothes

And the front door banged and I saw her
Walking crookedly, like naked, to a car.

She was not always with him up there,
And yet they seemed inviolate, like us,
Our loves in sympathy. Her going

Thrills and frightens us. We come awake
And talk excitedly about ourselves, like guests.

King and Queen

They are taming children in the garden flat.
We sit bolt upright as the guardians
Of a tomb. We are kingly with impotence.
Our arms lie along our knees. We might be
Ravelling wool with hollow hands.

Sugar Daddy

You do not look like me. I'm glad
England failed to colonize
Those black orchid eyes
With blue, the colour of sun-blindness.

Your eyes came straight to you
From your mother's Martinique
Great-grandmother. They look at me
Across this wide Atlantic

With an inborn feeling for my weaknesses.
Like loveletters, your little phoney grins
Come always just too late
To reward my passionate clowning.

I am here to be nice, clap hands, reflect
Your tolerance. I know what I'm for.
When you come home fifteen years from now
Saying you've smashed my car,

I'll feel the same. I'm blood brother,
Sugar daddy, millionaire to you.
I want to buy you things.

I bought a garish humming top
And climbed into your pen like an ape
And pumped it till it screeched for you,
Hungry for thanks. Your lip

Trembled and you cried. You didn't need
My sinister grenade, something
Pushed out of focus at you, swaying
Violently. You owned it anyway

And the whole world it came from.
It was then I knew
I could only take things from you from now on.

I was the White Hunter,
Bearing cheap mirrors for the Chief.
You saw the giving-look coagulate in my eyes
And panicked for the trees.

Family

In the bosom of the family
A court is in session.

The jury retire.
They run screaming through the streets.

Gone Away

We leave each other and the habits
Fall away like sight of land.

Now I am featureless
And you are infinite again.

Southampton

There's no one in the bar, so the barman
Looks at me and I remember him.

Was it really today
We crowded these sealed windows

Looking for the *Pasteur* in the pink-
Shot waters off Spithead?

The Elephant is Overturned

1

The elephant is overturned
And the snowstorm smashed.

The farm is scattered
And the mini-bricks

Lie old and dazzling
As jewels on the linoleum.

I let things lie.
It seems like donkey's years

Since my family went home.
From where they are

I think they look at me with love
And wonder why my future doesn't take.

2

Some hand
Has cut a section through this house.

Our bedroom is an open dig where we are petrified,
Naked as the lovers of Pompei,

Three thousand years
Beyond the day we walked in there,

Perhaps passionately, perhaps
For fun. Now we shall never know.

Night Club

I watch the ragtime couple
Throw their shoulders in the air,
And you are not here.

'I am in a little saloon
Drinking lemon water
And thinking of many things
And I am sorry . . .'

But what's the good of that my love,
If we are strung out like runners,
Losing ground?

I watch the ragtime couple
Throw their arms about each other,
And you are in Germany
Speaking to journalists.

'I remember my mother,
My little daughter . . .'

And I remember you. Your body
Furled in dusty Portobello weeds,
Your tragi-comic curls.

In Egypt

Torquoise wings on Diamond Harbour Road,
Miss Supra Bhose on Musky Street.
At Karnak, Gouda said 'That will be colossal.'
Cracking sunflower seeds
He told us about his lost master.

Sonny Jim

In my jacket and your jewels
The pusher is always with us.
Our little Sonny Jim.
We can't take our eyes off him.

I came home one night
And he was combing his hair like a mermaid.
I hung my coat on a peg
And it looked to me like a shroud.

Jim smiled lovingly round at me,
Long teeth in a skull. I thought:
'Some secret has dissolved his eyes.'
He bridled yellowly.

Since then we take him everywhere we go.
He is a monkey on a stick.
He only talks about one thing:
Sonny Jim is his own latest trick.

The prettiest boy in his class,
He can tell you all there is
To being a speedfreak at fifteen.
He has the track on his groin.

'They make you roll up your sleeves
And your trousers. I've seen jammed
Forearms come away like plaster
When a bandage is unwound.'

We rock him in our arms
And murmur at the world we bailed him from.
Moses in the bulrushes
Was not more loved than our Sonny Jim.

Sonny Jim Fears

His Carnaby Street frock-coat reminds him of himself.
Flared trousers hold up the dead courtier.

His ghost has sucked him into its vacuum.
He's naked in there, getting smaller.

*

They're going to garrotte him again.
He's buzzing with horror.
The human honeycomb is trying to replace itself.
The hive is empty.
Bees fly out of his ears.

*

The spare-part man is breaking up. Only his clothes
Hide the year-old accident
Looming like a becalmed hulk in his eyes.

Reconstituted from memory, he is not an accurate
Model of himself. His weathered skull
Can barely hold on to his hair.

The little faults are wearing holes in his silk skin.
He has had to close down whole areas of pain,
Like disused wings of a childhood.

If he moves at all, the atoms fizz and scream at him
And some burst spluttering over the top.
His life is something from a schoolboy's science kit.

The antidote is too far back for him to know.
He's fossilized, crumbling chalkily in space.
The little bits of him are wild geese.

In-Patient

I can find no argument, or even
Dumb insolence to comfort me
In this empty house like an aquarium.
Everything is seen to. They bring me tea

At all the right times, people
From under the stairs with rosy cheeks
And curls, whose steps are so gentle
You never hear them unless a board creaks

Or a door slams. I don't know why,
But after lunch they put a deckchair under a lime
And want me to sit here breathing quietly
Until I hear a bell go in the compound . . .

Travelling

Road soft as hair,
My shoulder brushes the avenue.
My head might nod to sleep
In the leaves which batter like windsails.
My eyes might reap the long fields.
I would be buried in the clouds.

February the 20th Street

A coincidence must be
Part of a whole chain
Whose links are unknown to me.

I feel them round me
Everywhere I go: in queues,
In trains, under bridges,

People, or coincidences, flukes
Of logic which fail
Because of me, because

We move singly through streets,
The last of some sad species,
Pacing the floors of zoos,

Our luck homing forever
Backward through grasses
To the brink of another time.

Some Sweet Day

Lone Star

A child of six who likes the cold wind
Stands in the garden, wondering
What to do now the others are in bed.

From her trapeze
She looks round at another day
Deserting her with promises.

Smudged by the setting sun,
Her cowboy T-shirt is illegible.
She faces into the wind,
Prepared to ride out this last day.

Pay Day

'At the end of the world,' said the crone
'Are blue and red days
Wrapped and waiting in the shade
And yours for the taking.'

I set one foot on the porch.
The boards creaked. The crone
Looked up from her sleeping with a start.
'The chillies are under the cloth,' she said
'But they ain't been paid for.'

Last Night

Wrapped in two dragons with red tongues
You are putting on
Blue eye shadow for work.
When you get to your feet
Your eyes will be closed again.
You shake some keys or light a cigarette.

I peer into the morning light
But you have disappeared
Into the wings of last night.
Your photograph of Blondin,
Tucked in a mirror,
Hovers between two days.

Money

The tulips press excited faces to the window
Where we are shouting about money.
The banks are closed and we are hungry,
Cut off from our roots like hothouse blooms,
Condemned to holiday
Behind these bars of sunlight
Striping the kitchen where we sit and smoke.
Weakened with too much coffee, we allow
The hours to gather us.
Time is for sale. We toil and spin
To stay on the same spot in the sun,
But money owns the day—unravelling
Our progress like a tapestry.
Slumped in our chairs,
We blame the dying clock
Which claimed to have our interests at heart.
A grim grandfather from Carmarthen,
It even has my name on it:
H.D.A. Williams Esq., 1910
And a slit for the phases of the moon.
Each quarter hour
It strikes—sealing the cracks in this cage.
The tulips shake their heads at us.
Like visitors to a menagerie they come
To shiver slightly at the latest specimen.

Sonny Jim Crowned

The haemophiliac boy king
Has blood veins growing in his hair.
He wears it wrapped in frozen towels
And always stays indoors
For fear the bleeding starts again.

They say we have to humour him
Because he's dying, so we stand around
Watching him point his toes in the long mirror,
Telling him how good he is at Ludo
In case he dies before we let him win again.

We're not allowed to mention the bits of blood
Which fly off him, staining our clothes,
When he skips among us like a little doll,
Showing off as usual.

Expatriates

The vineyard where we live
The one we draw about us on summer nights
Has influenced our poetry
The way it flavours its watermelons with silence.

Stampede

Clouds swimming in the paddies.
White peacocks pecking at their own reflections.
Sober as temples, the water buffalo
Are knee-deep in centuries.

From underneath their feet,
From underneath the clouds they are standing in,
A ripple is spreading
Which will muddy the stars.

Legend

They say the clouds are men and women
Fleeing together from the stampede
Their feet slowed down in our dreams.

We saw them for a moment once
And heard their cries
But they slipped back under the hooves
And were lost to the everlasting sky.

Tavistock Square

The Hiroshima silver birch 1963
Is not silver in January.

Mahatma Gandhi looks cold
On his stone shell like a loudspeaker.

They sit down facing him in silence
To talk about their life.

Husband and wife,
Why can't they believe in it?

Holidays

We spread our things on the sand
In front of the hotel
And sit for hours on end
Like merchants under parasols
Our thoughts following the steamers
In convoy across the bay
While far away
Our holidays look back at us in surprise
From fishing boats and fairs
Or wherever they were going then
In their seaweed headresses.

Lives

The same train each night
Enters the same station a little late
And different passengers going home
Look up from different books with a start
Their eyes narrowing back into a world
Hardly more shared
As they rake the length of the platform
For something they should find there
The thread of their lives
Difficult to recognize in the gloom.

Low Tide

Up there near the ceiling of our room
Is the high water mark.

Our dreams have fallen away from us.
We were almost real.

Truce

Each dawn might be winter again—
White empty skies
And all around the block
The curtains white.

Upstairs
We sleep where we are told to sleep.
Without this truce each day
God knows where we would be.

Empires

When I stare for hours
At the giant purple weeds
Wandering aimlessly over the battlefield
Of this garden, their airborne seeds
Spiralling up over the graves
Of the chrysanthemums
And suddenly decide to clear it all away,
Make it civilized here,
But go on staring at it, hypnotized by the task,
I think my stare
Is Nature looking back at me
With designs on my poetry.

The Nestling

Bird who stoops on my sleep,
Lifting me to the ledge
Where your open-mouthed young are waiting,
Tell me what I am.

In the morning I see a body like my own
Dangling from your claws
When I wake, expecting to be fed.

The North

To Neil Rennie

Water turns back
From the slow climb north.
Days turn back like children from the dark.

Snow stops falling
On fires, except by the shore.
We are here to see for ourselves
How it covers our footprints with the tracks of deer.

The Water Bearer

When I walk in the streets at night
Following the lamplight to where it falls
Exhausted in my head, some girl
Still carries my love on her shoulders through the crowd
Which sometimes offers up her face to me
Like a book which flickers shut again.

So long I have tried to touch that face
When it drifts for a moment near me on the tide.
So many times I have seen it
Sucked back into the sea
Of all such nights I follow down like stones
To where they lie unfound, unfathomable.

When I wake in the morning, far from her,
This girl, wherever she is sleeping, wakes with me
And takes up the weight of my love for her,
Carrying it back into the world of absences
Where I see her walking alone in the streets
Cursed as she is with being mine.

For her shoulders that are always disappearing
Into the heart of her own searching
Look so calm and beautiful despite it all
That I wait impatiently for the sight of her
Passing again so sweetly through my life
As if she carried water from a well.

Li-Lo

When the rain touches on my past
I lie back to the earth.
The vine, the wine, the wilderness—
I'm seeing stars!

Rain on my drunken face, the drops
Are timed to sink my thoughts of you.
But they cool my skin,
Is that coincidence?

Century Oaks

The trees are emptying.
The cold young days rush through them
On their way to power.

Down here
We sweep the dead leaves into bonfires
Lest they betray our sympathies.

Cherry Blossom

For Soph Behrens

Delicate red flames are finding their way
Out of the dark cherry branches.

Already they expect the clouds
To reward their fidelity.

Do they come out of the dragon earth,
Or out of my mouth as I had feared?

I see their torches searching my mind
For a clue to their mystery.

They find instead some other little man,
Wandering the hillside with an unknown speech,

Gesturing wildly at the cherry trees.
Are we all the same?

And why are we at war
With easy things we are the reason for?

The cherry-blossom flowers
Wither away from me with a sigh,

As if a great ship were getting under way,
Its coloured streamers

Lifting and breaking finally
As it pulls away from the quay.

I wish the rain could quench this fever,
Which is like an image going on forever.

But rain is not rain to me any more,
It is a warrior at the window pane.

I stand convicted by the laws of nature
Of gazing too long into the future.

O lovely cherry tree, come true.
Together we could celebrate the years.

Madonna

On the stairs to the dormitories
I used to go past
A picture of the Virgin Mary
Made out of butterfly wings
Which cast a glow
Over the gloomy photographs
Of Field Marshal Earl Haig
And Admiral Jellicoe.
She has cast her friendly beams
Into the wilderness of my life.
I salute her in her solitude.

To His Daughter

A little girl sits on the wall
Colouring something on the other side.
The wind shines her hair as she holds up
A flapping paper to my window.

What is on the paper I don't know.
The hornbeam bends over near her head.
Its two-coloured leaves are like her hair.

Prayer While Sleeping

Brave blanket covering me
With a theme of arrows against the night,
Draw the coloured curtain tight,
Look after time for me.

Keep him satisfied while I sleep,
Slung like a bridge over this gorge,
This no-man's-land.
Keep him busy with his sand,
For I owe him emptiness.

Love-Life

Now and Again

We lift the bamboo curtain to the old hothouse.
We see the upturned faces of our guests.
They are smiling patiently
As if they have been expecting us.
They sign to us to approach.

It is now that your veil blows across my face
As we move towards the table for the feast.
We cut the cake like this,
Holding the knife and wishing for different things.
They drink to our happiness.

Bar Italia

(To L.S.)

This is how we met,
Sheltering from work in this crowded coffee bar.
This is where we sit,
Propped at some narrow shelf,
Each day more crowded than the last
With undesirables.

I wish we could meet again
In two years' time,
Somewhere expensive where they remembered us
From the early days, before the crash.
Instead of here, instead of now,
Facing our reflections in the Bar Italia.

You would be frowning of course,
After all this time,
But then you would hold out your hand and say,
'Well, where are your three pages?'

I haven't written them. One day I will.
Anywhere but here it might seem possible.

Kites

Our lives fly well—white specks with faces
Running out against blue. While far below
We stand staring after them,
Trying to remember what they were like,
These prize possessions of ours,
Unravelling so cheerfully before our eyes.

By now we are winding in the runaway spools
For all we are worth. Whatever was there
Has begun to recede, like the dead stars,
Faster than the speed of their light
Can reach back to us here,
Where we hang on these empty strings.

Tides

The evening advances, then withdraws again
Leaving our cups and books like islands on the floor.
We are drifting you and I,
As far from one another as the young heroes
Of these two novels we have just laid down.
For that is happiness: to wander alone
Surrounded by the same moon, whose tides remind us
of ourselves,
Our distances, and what we leave behind.
The lamp left on, the curtains letting in the light.
These things were promises. No doubt we will come back
to them.

Impotence

You see me with my suits, my well-cut suits:
Past, present and future
Ranged close at hand upon their hooks.

How you hated them, hanging there so like me they hurt,
The herringbones, the faint chalk stripes,
Withdrawn from the wear and tear.

They were never in the wrong, the multi-pocketed,
The stay-at-homes. They were cosy-warm,
Huddling together there.

But where did you go that night
While I hung about upstairs, unwell, unable to decide,
Should I wear this one? Or that?
Take off the tie, or keep the waistcoat on?

You couldn't wait
When time ran out on me. I crossed the floor too late
To shut the cupboard which contained the sea.

Bachelors

What do they know of love
These men who have never been married?
What do they know
About living face to face with happiness
These amateurs of passion?
Do they imagine it's like home used to be,
Having a family of one's own,
Watching the little bones grow lethal,
The eyes turned on you—
And realizing suddenly that it's all
Your own fault the way things are,
Because it's you now
Not your parents who're in charge?
Can they understand what it means,
These suntanned single men? Or are they into cars?

And what do they know about the bedside lamp,
These denimed Romeos,
Its sphere of influence as night descends.
Familiar switch to hand:
On-off, off-on, the thousand little clicks
Half in, half out of the dark,
As the row gets going on time, or nothing does,
Or the bulb just sings to itself
On your side of the bed?
Pride in anger. That's your happiness.
A poisonous seed washed up with you
On a desert island of your own making,
Your impotence in flower like a hothouse rose.
And they talk about love
These men who have never been married.

Come, Tears

The days are full of darkness, the evenings glow
With longing for a year
That's going out. Come, night,
Fall fast on this house. Let me sleep
In love with her again. Come, tears, and fall
For other nights like this,
Whose spell I broke by so deceiving her.

Rain hissing at the window, the world is stuck
In the last groove of a year
That's nearly gone. Come, rain,
Splash your brushstrokes down
Grey stone. Come, tears, why don't you fall like that
On my hands, that I may find
This sadness no harder to bear?

A New Page

I write your name on a new sheet
To see if it will stand my weight.
It zig-zags like a crack
On the frozen surface of a lake
Which will only bear me up so long
As I keep moving.

A thousand unseen weaknesses
Turn foggy as they pluck like kisses
At thin ice. Uncertain territory
Stretches ahead, while behind me
A delta of tiny breakages begins
To scatter its reasons
Why I never can go back—
These words collapsing in my wake.

I should know where I am
By now in this pocket snowstorm
Shaken by a child, whose hands hold
A little snow-filled world
Up to the light
To see me struggling to stay upright
In her story. But I am lost
In a patch of weather from the past.

Familiar shadows queue
To usher in old memories, while through
The mist I see your ice-locked eyes
Looking back at me from these
Same sentences. I wonder will
They ever tire of their vigil.

Their longed-for colours of tenderness,
Changing to those of glass
Without warning, are still known only
To themselves, whose treachery
Was my own. Now that the page
Is covered with words in your image,
Will your heart melt? Will you break
The ice between us, for my sake?

Your Way Home

I turn my back on you and have to watch the cars
Competing on the rain-washed motorway. Only the rain
And a line of poplar trees which seems to end
At this rustic slum of ours can point
To why we huddle here, in the middle-distance, shivering.

When I shout at you to leave me on my own
You know very well what I mean, but you don't come near.
For once you just do as I say—your suitcases
Clenched tight round property and the past. I could weep
For showing so little strength with you.

In the lull, I can hear the poplar trees
Pouring on wind such ecstasy as we have known
In this old barn with beds for furniture,
Hearing the rain on the sloping roof all night,
Not hoping it would stop or let us go.

Tonight the same night is drawing in its colours, now
As then, though we of that night are not the same.
And nature can do nothing more for us this time
Than count the hours till you must find your way
Back from this place and start to live again.

Your absence falls in front of you. Your gestures have
  withdrawn
Almost to the horizon, where the rain, filed thin,
Floats in across the miles of flood-fed earth
That will come between us. I can see your car
Move up the slip-road into the London lane.

Broken Dreams

The women sleep.
We look for them in their dreams.

When we bump into a piece of the scenery,
It falls, waking them.

They open eyes full of broken love.
Love that we have broken.

The White Hair

The hungry hours of the earth grope through me now
In their search for images.
I wish I could pluck you out of me
As easily as the white hair I saw in the mirror,
Though even then I noticed my searching right hand
Start moving in the wrong direction.

Along These Lines

And so you cry for her, and the poem falls to the page
As if it knew all along that what we make of ourselves we
take
From one another's hearts—tearing and shouting until we
learn
How awkwardly, upstairs and behind shut doors we are born
Already owing interest on what we have borrowed from the
world.

The Ribbon

I thought she'd taken all her things
But I was wrong. Wherever I go
I catch glimpses of my damnation.

Is that too strong a word? You wouldn't think so
If you could see this lovely ribbon
Wound around my hand.

Don't tell me, I know,
I'm mumbling to myself again. I'm like King Kong
Picking among the ruins of New York
For a clue to his misfortune.

I keep wondering what they were like,
These odds and ends,
Collecting dust, though freed at last from blame.
Did they look the same
When she held them in her hand?

It seems ridiculous
How everything here acknowledges her touch,
Including me, including this tangled ribbon.

Perhaps you were right after all
And I make too much of it. I'll just sit here now
And try to undo these knots—
I'll be with you in a minute, if you can wait.

Stagefright

(*thinking of my father*)

Dazed with the sadness of lost things
In ordered silence
I sat down in the dining room for tea:
Biscuits and a glass of Moselle, no radio.
(How kind we are to ourselves!)
And I tried to imagine
What he would have done at a time like this
For I will say this
He knew what to do in an emergency
And he knew what to drink.
So I put some more wine in the fridge
And I hurried round to her house.
I shouted her name and knocked
But she spoke to me through the frosted glass
And I'm glad I couldn't see her face
When she told me it was all over between us.

Then I shivered like a man with stagefright
And I watched the world
Come slowly to a standstill before my eyes:
The sinking of the heart.
It seemed unacceptable suddenly
To be walking the streets on such a night
With love like so much small change
Left over from a pound.
I could hear him telling me:
'Women are strong, but they fall
Like sleep from your eyes. Let your step
Spring on the pavement and you'll see
Your only fault's unhappiness.'
Then I came back here and got into my good suit,
Having chucked the biscuits
And opened the bottle of wine.

Present Continuous

Well, I am still
The unofficial guardian of your house,
Which is not your house any more
And not the same place we trusted to be there
Whenever we came home.
Our possessions lie
Abandoned, back along the way:
These books, those dresses under cellophane.
I haven't moved
Your plastic carrier bags from the hall
And fifty pairs of shoes
Still hang around the window on the stairs,
The changing fashions of your years with me.

Late

When the record slides to a close
I imagine your key in the lock
And your handlebars bumping against the wall
You are so angry to be back.

Look—you have still got me
Gliding about your business as before.
Wherever you are out there, in the cold night air,
You would be proud of me.
I've never been so good.

After the Show

It's been too long since I waited up for you
With something to eat after the show
And everything done.
Too long since I sat here with a grin
While the bloody show moved on.

As if to remind me,
The news comes round again
About a famous ballet star's defection.
By midnight, she is blond and beautiful
And there is a man in her life.

Are you blond and beautiful at last?
Or don't you care?
When people ask after you I have no news for them
Except that you are far away from here
And everything is forgiven.
I wonder how they went, those simple nights,
Before the years set in.

Love at Night

She is with me now, my lifelong visitor,
She who has faithfully instructed me
In the art of being alone. She will not rest
Till she has taught me all there is to love
And left me here to learn the facts by heart.

Even now, she is undressing in her mirror
While I lie here, awaiting tonight's lesson
On how to exist without her another hour.
I know what she will say: 'Don't look at me.
I feel guilty coming here like this
After so long. I don't know what to do.'

See how gently she has persuaded me
That every night I spend with her is my last.
Merciless angel, it has been your task
To teach me how to live without you finally.

Confessions of a Drifter

I used to sell perfume in the New Towns.
I was popular in the saloons.
Professional women slept in my trailer.
Young salesgirls broke my heart. For ten years
I never went near our Main Office.

From shop to shop
And then from door to door I went
In a slowly diminishing circle of enchantment
With 'Soir de Paris' and 'Flower of the Orient'.

I used up all my good luck
Wetting the wrists of teenagers in bars
With 'English Rose' and 'Afro-Dizziac'
From giveaway dispensers.

From girl to girl
And then from bar to bar I went
In a slowly expanding circle
Of liquid replenishment.

I would park my trailer outside a door
So I could find it when I walked out of there,
Throwing back my shoulders at the night—a hero
To myself.

They knock on my window this morning. Too late
I wake out of my salesman's paradise,
The sperm drying on my thigh
And nothing but the name of a drifter in the New Towns

Love-Life

Her veil blows across my face
As we cling together in the porch.
Propped on the mantelpiece,
The photograph distils our ecstasy.
Each night we touch
The heart-shaped frame of our reliquary
And sigh for love.

Each morning we are young again—
Our cheeks brushed pink,
The highlights in our hair.
Our guests will be arriving soon.
We wait contentedly beyond the glass
For them to find us here,
Our smiles wrapped in lace.

Writing Home

At Least a Hundred Words

What shall we say in our letters home?
That we're perfectly all right?
That we stand on the playground with red faces
and our hair sticking up?
That we give people Chinese burns?
Mr Ray, standing in the entrance to the lavatories
with his clip-board and pen,
turned us round by our heads
and gave us a boot up the arse.
We can't put that in our letters home
because Mr Ray is taking letter-writing.
He sits in his master's chair
winding the propeller of his balsa wood aeroplane
with a glue-caked index finger
and looking straight ahead.
RESULTS OF THE MATCH, DESCRIPTION OF THE FLOODS,
THE LECTURE ON KENYA, UGANDA AND TANGANYIKA
WITH COLOUR SLIDES AND HEADDRESSES.
We have to write at least a hundred words
to the satisfaction of Mr Ray
before we can go in to tea,
so I put up my hand to ask if we count the 'ands'.
Mr Ray lets go the propeller of his Prestwick 'Pioneer'
and it unwinds with a long drawn-out sigh.
He'd rather be out overflying
enemy territory on remote
than 'ministering to the natives' in backward C4.
He was shot down in World War One or World War
Two, he forgets,
but it didn't do him a damn bit of harm.
It made a man of him.
He goes and stands in the corner near the door
and offers up his usual prayer:
'One two three four five six seven
God give me strength to carry on.'
While his back is turned
I roll a marble along the groove in the top of my desk
till it drops through the inkwell

on to the track I've made for it inside. I can hear it
travelling round the system of books
and rulers: a tip-balance, then a spiral,
then a thirty-year gap as it falls through
the dust-hole into my waiting hand.

Just Another Day

When you were young
you came downstairs in the middle of the night
and saw the living room.
The furniture lay about your feet.
The carpet had been folded back
where it met the skirting board.

You opened the front door
and stood for a moment on the step.
Little pieces of metal
shone in the asphalt on the road.
The chimneys were pot-bellied apostles
preaching to the stars.

You cleared your throat, or coughed,
and the dawn chorus started up—
excited by an item of news
which might have been you,
or might have been just another day.
You stood there for a moment, listening.

A Walking Gentleman

I started very slowly,
being rude to everybody
and going home early
without really knowing why.
I carried on that way
till my father died
and allowed me to grow my hair.
I didn't want to any more.
I came through a side door,
my hands slightly raised,
as if whatever was going on
needed lifting by me.
I bought a clove carnation
in Moyses Stevens
and walked all the way up Piccadilly
to the top of the Haymarket,
stopping every so often.
Surely Scott's is somewhere near here?
I can't see it any more.
My feet are hurting me.

Before the War

'You should have been there then,' they tell you,
the girls who were there themselves.
'Before the war,
your father was the kind of man
to take you, on the spur of a telegram,
to one of those Continental casinos
where they keep the curtains drawn
all summer: white ties and Sidney Bechet,
gardenias on a breakfast tray.
You'd follow the road map south
in someone's aeroplane,
putting down in a field while it was light.
Oh, those were the days all right
and the nights too for someone like your father.'

Then you mourn the fact once more
that you missed knowing him then,
that you hardly recognize this man
who somehow jumped the gun
and started ahead of you. It isn't fair,
but there's nothing to be done. The casinos are dead
and the nights are drawing in.
Though you follow the road map south
on the spur of a lifetime
you'll never catch up with the fun
and he won't be back for you.
You're strung out like runners
across the world, losing ground,
in a race that began when you were born.

Waiting To Go On

I turned the pages slowly, listening for the car,
till my father was young again, a soldier,
or throwing back his head
on slicked back Derby Days before the war.
I stared at all that fame and handsomeness
and thought they were the same.
Good looks were everything where I came from.
They made you laugh. They made you have a tan.
They made you speak with conviction.
'Such a nice young man!' my mother used to say.
'So good looking!' Day after day
I searched my face for signs of excellence,
turning up my collar in the long mirror on the stairs
and flourishing a dress sword at myself:
'Hugh Williams, even more handsome in Regency!'
The sound of wheels on the drive
meant I had about one minute
to put everything back where I'd found it
and come downstairs as myself.

Tipping My Chair

I shivered in 1958. I caught a glimpse
of money working and I shut my eyes.
I was a love-sick crammer-candidate, reading
poetry under the desk in History,
wondering how to go about my life.
'Write a novel!' said my father.
'Put everything in! Sell the film rights for a fortune!
Sit up straight!' I sat there, filleting
a chestnut leaf in my lap, not listening.
I wanted to do nothing, urgently.

At his desk, in his dressing-gown,
among compliant womenfolk, he seemed
too masterful, too horrified by me.
He banged the table if I tipped my chair.
He couldn't stand my hair. One day,
struggling with a chestnut leaf, I fell over backwards
or the chair-leg broke. I didn't care any more
if poetry was easier than prose. I lay there
in the ruins of a perfectly good chair
and opened my eyes. I knew what I didn't want to do.

At his desk, in his dressing-room, among
these photographs of my father in costume,
I wonder how to go about his life.
Put everything in? The bankruptcy? The hell?
The little cork-and-leather theatrical
'lifts' he used to wear? The blacking for his hair?
Or again: leave everything out? Do nothing,
tip my chair back and stare at him for once,
my lip trembling at forty?
My father bangs the table: 'Sit up straight!'

An Actor's War

Tunisia 1943

> Before the British public
> I was once a leading man.
> Now behind a British private
> I just follow, if I can.
> —Hugh Williams

March

Well, here we are in our Tropical Kit—
shirts and shorts and little black toques,
looking like a lot of hikers or cyclists
with dead bluebells on the handlebars.
It seems we have at last discovered a place
where it is impossible to spend money. What a pity
that it should be a rather muddy wadi
in Tunisia, where whisky is prohibited by God.
How sorry I am that I ever said an unkind word
about the Palmer's Arms. In my nostalgia
it seems the very Elysium of Alcohol.
I can imagine you in about an hour
pattering round to meet your beaux.
The last couple of days I've realized with a bang
what an appalling time this bloody war has been on.
Three and a half years last night
since we walked out of the stage door of the Queen's Theatre
into the Queen's Westminsters.
What good times we had. But it all seems
a long time ago, looking back, doesn't it?

April

Early morning—or what in happier times
was late at night. Strong and sweet black coffee,
laced with the last little drop out of my flask,
has reminded me of that stuff they used to serve
on fire inside a coconut at The Beachcomber
to put the finishing touches to a Zombie.
I'm still floundering in the work here.

I lie awake sometimes wondering if my map
is marked correctly. I lose notebooks
and have to rely on little bits of paper.
Benzedrine tablets, please. Chemist next to the Pavilion.
A kiss and a lump of chocolate for Hugo
for being able to walk.
Please God he never has to march.

May

It's all very green down here at the moment—
lots of wild flowers and lots of your gum trees
with their barks hanging down like tattered lingerie.
I saw a stork flying and heard a lark singing
as though he were over Goodwood racecourse
on that wonderful day when Epigram won the Cup
and you won me. The villages look like those
in Provence and the milestones with little red tops
make me long for the days to come
when you and I are scuttling down the Route Bleue
*in search of sunshine and eights and nines.**
Having taken trouble all one's life to seek pleasure,
to find now that delights are down to a canvas bath
taken with one's legs hanging over the side in a bucket,
is strange, though no doubt good for one.
I dare say I shall be pretty bloody exquisite
for quite some time after the war—silks and lotions
and long sessions at the barber
and never again will a red carnation be made to last
from lunchtime until the following dawn.
When the war is over I intend no longer
to practise this foolish and half-hearted method
of letting money slip through my fingers.
I intend in future to allow it to pour
in great torrents from my pockets.
Don't be alarmed. This is only the talk of a man
with mosquito lotion on his face and hands
and anti-louse powder in the seams of his clothes,
who drinks his highly-medicated morning tea

* the good cards in Chemmy

from a tin mug with shaving soap round the rim
and uses gum boots for bedroom slippers.

June

Writing by our Mediterranean now, but the wrong bank.
The same sunshine and azure sea, a few of the same
flowers and trees and the purple bougainvillea,
but there it ends. Enough to make one want more—
a bottle cooling in a pool,
a yellow bathing dress drying on a rock.
Perhaps if we fight on we shall arrive in a country
where there is something fit to drink.
How pleasant to be advancing through the Côte d'Or
with one's water bottle filled with Pouilly.
Instead of which we're stuck in this blasted cork forest
learning to kill flies.
Sometimes it seems we love England
more than each other, the things we do for her.
I wonder if, when it's over, we'll be glad.
Or shall we think I was a fool to sacrifice so much?
Oh God, we'll be glad, won't we? I don't know.
Not in this damned dust hurricane I don't.
But if you love me I shan't care.
You and Hugo have a coating of desert on your faces.
I must wipe you.

July

The battle—if one can dignify such a shambles—
is closed in this sector and there is an atmosphere
of emptying the ashtrays and counting the broken glasses.
Churchill arrived to address the First Army
in the Roman Amphitheatre at Carthage.
He looked like a Disney or Beatrix Potter creature
and spoke without his teeth. Cigar, V-sign, all the tricks,
and I thought of that day outside the Palace
with Chamberlain smiling peace with honour
and we kidded ourselves there was a chance—
two little suckers so in love
and so longing for a tranquil sunny life.

August

How's my boy? Shirts and trousers!
Poor little Hawes and Curtis. Another year or so
and our accounts will be getting muddled
and I shall find myself getting involved
in white waistcoats I've never seen.
Tell him to pay cash. Go and tell him now.
The thought terrifies me.
Have been harassed lately by the old divided duties—
the only part of the war I can honestly say
has been bloody. Maybe the cinema racket
gives one the wrong impression of one's worth,
but I sometimes feel I'd be better employed at Denham
as Captain During RN than housekeeping for Phantom.
Stupid, for one must do one or the other
and not attempt both as I have done.
Had a letter from the Income Tax
asking for some quite ridiculous sum.
Next time you see Lil tell her to write and say
I'm unlikely to be traceable
until quite some time after the war, if then.
I think when I die I should like my ashes
blown through the keyhole of the Treasury
in lieu of further payments.
My wages here are roughly what it used to cost me
to look after my top hat before the war.
Flog it, by all means. I can't see that kind of thing
being any use after the war, unless it's for comedy.
Did some Shakespeare at the Hospital Concert
the other night and was nervous as a cat.
God knows what a London first night will be like
with all the knockers out front, waiting and hoping.
I doubt if I'll make it. Sometimes I really doubt it.
I'll probably run screaming from the theatre
just as they call the first quarter.
Tell the girls to keep on with Puck and the First Fairy
as I shall want to see it when I come home.

September

Had a deadly exercise down on the plain last week
and the blasted Arabs stole my lavatory seat.

Medals should be given for exercises, not campaigns.
One would have the Spartan Star for Needless Discomfort
in the face of Overwhelming Boredom.
I had to give a cheque for £48 to Peter Baker
and I doubt there's that much in my account.
Now he's going home by air because of an appendix
and taking the cheque with him.
I couldn't be sorrier to do this to you once again,
but his appendix took me by surprise, as it did him.
Tell Connie I must have a picture before Christmas.

December

Every known kind of delay and disappointment
has attended us and I am filled with a sulky despair
and a general loathing for mankind.
People are so bored they have started growing
and shaving off moustaches, a sure sign
of utter moral decay. I have luckily made friends
with a little fellow who keeps me supplied
with a sufficiency of Algerian brandy,
so I expect the major part of my waking life
to be spent in pain and hangover.
Added to all other horrors,
Christmas Theatricals have cropped up,
which really has crowned my ultimate unhappiness.
Perhaps if I tell you that after
an hour and a half of forceful argument
I have just succeeded in squashing an idea
to produce an abbreviated version of Midsummer Night's Dream
by the end of the week—without wigs, costumes,
stage or lighting and only one copy of the play,
you will appreciate the nervous exhaustion I suffer.
The idea of acting is rich. Not for a line of this letter
have I avoided making those aimless
slightly crazy-looking gestures to remove the flies.
I have a mug of tea and there must be thirty round the brim.
I can kill them now by flicking them,
as opposed to banging oneself all over.
I think they must be slower down here,
for I can't believe that I am quicker.

Tangerines

'Before the war' was once-upon-a-time
by 1947. I had to peer through cigarette smoke
to see my parents in black and white
lounging on zebra skins, while doormen stood by doors
in pale grey uniforms.

I wished I was alive before the war
when Tony and Mike rode their bicycles into the lake,
but after the war was where I had to stay,
upstairs in the nursery, with Nanny
and the rocking-horse. It sounded more fun
to dance all night and fly to France for breakfast.
But after the war I had to go to bed.

In my prisoner's pyjamas I looked through
banisters into that polished, pre-war place
where my parents lived. If I leaned out
I could see the elephant's foot
tortured with shooting sticks
and a round mirror which filled from time to time
with hats and coats and shouts,
then emptied like a bath.

Every summer my parents got in the car
and drove back through the war to the South of France.
I longed to go with them, but I was stuck
in 1948 with Nanny Monkenbeck.

They sent me sword-shaped eucalyptus leaves
and purple, pre-war flowers, pressed
between the pages of my first letters. One year
a box of tangerines arrived for me from France.
I hid behind the sofa in my parents' bedroom,
eating my way south to join them.

Slow Train

My father let the leather window-strap
slip through his fingers and I smelt the sea.
He was showing me gun emplacements
to stop me feeling train-sick
on our first holiday after the war.
I clutched my new bucket in two lifeless hands,
excited by the blockhouse
which had exploded, killing everyone.
We went over a bridge he had guarded
and he lit two cigarettes and threw them down
to some soldiers cutting barbed wire.
He said there was something fast for me
in the guard's van, if I could hang on.
I sat there, staring at one of the holes
in the window-strap, imagining death
as a sort of surprise for men in uniform.
'I-think-I-can-I-think-I-can'
the train was supposed to be saying
as we came to Dungeness Lighthouse in the dark,
but I didn't think I could.
When we started going backwards, I was glad.

Now That I Hear Trains

Now that I hear trains
whistling out of Paddington on their way to Wales,
I like to think of him, as young as he was then,
running behind me along the sand,
holding my saddle steady
and launching me off on my own.

Now that I look unlike
the boy on the brand new bike
who wobbled away down the beach,
I hear him telling me: 'Keep pedalling, keep pedalling.'
When I looked over my shoulder
he was nowhere to be seen.

Walking Out Of The Room Backwards

Out of work at fifty, smoking fifty a day,
my father wore his sheepskin coat
and went to auditions
for the first time in his life.
I watched in horror from my bedroom window
as he missed the bus to London
in full view of the house opposite.
'If it weren't for you and the children,'
he told my mother from his bed,
'I'd never get up in the morning.'

He wasn't amused
when I burst in on his sleep
with a head hollowed out of a turnip
swinging from a broom. There were cigarette burns
like bullet-holes in his pyjamas.
I saw his bad foot
sticking out from under the bedclothes
because he was 'broke'
and I thought my father was dying.
I wanted to make him laugh, but I got it wrong
and only frightened myself.

The future stands behind us, holding ready
a chloroform-soaked handkerchief, in case we make a slip.
The past stretches ahead, into which we stare,
as into the eyes of our parents
on their wedding day—
shouting something from the crowd
or waving things on sticks
to make them look at us. To punish me,
or amuse his theatrical friends,
my father made me walk out of the room backwards,
bowing and saying, 'Goodnight, my liege.'

Making Friends With Ties

His khaki tie was perfectly knotted in wartime.
The tail was smartly plumped.
The dent became a groove
where it entered a sturdy, rectangular knot,
never a Windsor.

This groove came out
in exactly the same place all his life,
never in the middle,
but slightly to the left.
'You have to get it right first time,'
he told me, my first term at school.
'Otherwise you go raving mad.'

I was so impressed by this
I didn't listen in class.
I made friends with peoples' ties, not them.
One day when I was drunk I told him,
'I don't like the groove!'
His face softened towards me for a moment.
'Don't you, dear boy? Well, I'm *delighted*.'

Leaving School

I was eight when I set out into the world
wearing a grey flannel suit.
I had my own suitcase.
I thought it was going to be fun.
I wasn't listening
when everything was explained to us in the Library,
so the first night I didn't have any sheets.
The headmaster's wife told me
to think of the timetable as a game of 'Battleships'.
She found me walking around upstairs
wearing the wrong shoes.

I liked all the waiting we had to do at school,
but I didn't like the work.
I could only read certain things
which I'd read before, like the Billy Goat Gruff books,
but they didn't have them there.
They had the Beacon Series.
I said 'I don't know,'
then I started saying nothing.
Every day my name was read out
because I'd forgotten to hang something up.

I was so far away from home I used to forget things.
I forgot how to get undressed.
You're supposed to take off your shirt and vest
after you've put on your pyjama bottoms.
When the headmaster's wife came round for Inspection
I was fully dressed again, ready for bed.
She had my toothbrush in her hand
and she wanted to know why it was dry.
I was miles away, with my suitcase, leaving school.

Scratches

My mother scratched the soles of my shoes
to stop me slipping
when I went away to school.

I didn't think a few scratches
with a pair of scissors
was going to be enough.

I was walking on ice,
my arms stretched out.
I didn't know where I was going.

Her scratches soon disappeared
when I started sliding
down those polished corridors.

I slid into class.
I slid across the hall into the changing-room.
I never slipped up.

I learnt how to skate along with an aeroplane
or a car, looking ordinary,
pretending to have fun.

I learnt how long a run I needed
to carry me as far as the gym
in time for Assembly.

I turned as I went,
my arms stretched out to catch the door jamb
as I went flying past.

When Will His Stupid Head Remember?

Mr Ray stood behind me in History,
waiting for me to make a slip.
I had to write out the Kings and Queens
of England, in reverse order, with dates. I put,
'William I, 1087–1066'. I could smell the aeroplane glue
on his fingers as he took hold of my ear.
I stood in the corner near the insect case,
remembering my bike. I had the John Bull
Puncture Repair Kit in my pocket: glass paper,
rubber solution, patches, chalk and grater,
spare valves. I was 'riding dead'—
freewheeling downhill with my arms folded
and my eyes shut, looking Mr Ray in the eye.
Everytime I looked round he added a minute to my sentence.

Mr Ray held his red Biro Minor like a modelling knife
to write reports. He drew a wooden spoon.*
'I found it hard to keep my temper
with this feeble and incompetent creature.
He was always last to find his place
and most of his questions had been answered
five minutes before . . .' I called my father 'sir'
when he opened the envelope and shouted.
I was practising stage-falls from my bike
in the fading spotlight of summer lawns,
remembering the smell of aeroplane glue and inkwells
with a shiver down my spine. The beginning of term
was creeping up on me. Every time I looked round
Mr Ray was standing there, stockstill.

* 'bottom of the class'

Shelf Life

1
Above our beds
the little wooden shelf
with one support
was like a crucifix
offering up
its hairbrush, Bible,
family photograph
for trial by mockery.

We lay in its shadow
on summer nights,
denying everything,
hearing only
the impossible high catches*
for the older boys,
their famous surnames
calling them to glory.

2
Why did we take
the bed-making competition
so seriously?
We were only nine.
We measured our turndowns
with a ruler.
We used a protractor
to fix the angle of our
hospital corners
at forty-five degrees.

Our shelves were identical.
Our Bibles lay
on their sides, facing in.
Our hairbrushes lay on their backs
with a comb stuck in them.

* fielding practice

If anyone's hairbrush had a handle
they had to hide it
in their dressing-gown
and borrow a proper one
for the competition.

In the centre of our shelves
stood our photo-frame,
a difficult area
that couldn't be tidied away
or forgiven. By the time we had
solved the problem
of our counterpane
our parents were looking
straight past one another
into opposite corners of the room.

3
Their smiles were
lost on us
and ours on them,
as if they were still
waving goodbye to the wrong
upstairs window
from the car.
In their long absence,
our double photo-frame
was a bedtime
story-book,
propped open like a trap
at the pictures.

We said to ourselves,
'Brothers and sisters
have I none,
but this man's father
is my father's son.
Who am I?'—
holding our fingers
on our father's

encouraging smile
and repeating it
over to ourselves
till we started
to lose our place.

4
I knew it wasn't my father
who was bankrupt and poor.
He had a war.
He had a scar.
He was on Famous Film Star
Cigarette Cards
with Janet Gaynor.
It couldn't be my father
who hit the registrar
and had to be bound over for a year
to keep the peace,
so who were they talking about
in the newspaper?
If he was famous,
why hadn't I heard of him?

He looked uncertain
in the signed
photograph on my shelf
that was attracting
too much attention
for my own good.
His hair was perfection.
His eyes were fixed on the horizon
where something vaguely
troublesome was going on
behind my back.
The smoke from his cigarette
had been touched in
against a background
of pleated satin.

5
I found his name
in the Library *Who's Who*

and tore the page out
hoping it would say.
I memorized dozens
of forgotten films and plays
to prove my father
innocent of bankruptcy.
His brief biography
was followed by a personal note:
'Clubs: none, Sports: none,
Hobbies: none.
Address: c/o *Spotlight*.'

6
I tried to explain
that the German
bubble-car
in the photograph
of our house
was part of
a Spitfire
my father had flown
in the war.
The swastikas
on my blanket
were ancient
symbols of fortune
the other way round.

I sat in bed
tracing the faces
of my parents
on lavatory paper.
Riddles and smut
poured from their lips
in my defence,
but the evidence
was attached to
a blind-cord.
Up it flew,
hoisting my shit-
stained underpants
into full view.

Man Junior

I look over the banisters and see, far down,
Miss Pyke taking Roll Call. I push my feet
into a pair of Cambridge house shoes
half my size and shuffle downstairs.
When I answer my name there is a long silence,
then Miss Pyke asks me where I've been.
I tell her I was reading a book
and didn't notice the time.

I see I have a smaller desk this term
as a punishment for being late.
I have to sit sideways, facing Armitage,
who eats little pieces of blotting paper
dipped in ink. When the bell goes
I barge off down Lower Corridor
with my head down and my elbows out,
knocking everyone flying.

Hurrah! There's a letter for me today.
I'd rather have a parcel, but I'm always happy
when I see the familiar blue envelope
propped on the mantelpiece
on the other side of School Hall.
I don't open it straight away, of course.
I shove it in my pocket
and read it later, like a man.

I'm standing outside the Headmaster's Study
waiting for the green light to come on.
Either I've failed Common Entrance
or my parents have died. When I go in
he's sitting at his desk, staring out the window.
For a long time we watch Sgt Burrows
pushing his marker round Long Field,
Mr Harvey taking fielding practice.

The Headmaster pulls his writing case towards him
and opens it with his paper knife.
Inside is the worst news in the world,
my copy of Man Junior with a picture of a girl
in a bikini playing with a beach ball.
I must have left it under my mattress.
The Headmaster looks at me in disbelief
and asks, 'What is the meaning of this?'

There

If I got into trouble I was to go to him
and tell him everything.
It didn't matter if I was unhappy, or in love,
or wanted by the police. I could say,
'Daddy, I've killed a Chinaman'
and he'd see what he could do.

He gave me a knife for my birthday
and I cut my hand on it
and flung it away into the long grass,
running after it in vain
as it started to disappear.

It was easy to imagine myself
finding the knife and wearing it on a chain
the way he had shown me,
but not so easy when the grass had grown
and been cut many times,
the garden gone next door.

I must have been looking too far away
or I must have been looking too near.
A wealth of personal detail
accumulates in folders, like a life of crime,
but nothing conclusive,
nothing to get arrested for.

New Ground

We played Scrabble wrong for years.
We counted the Double and Triple Word Scores
as often as we liked.
We had to move aside the letters
to see what colours they were on.

My father was out of work
and we were moving again. He stared at the board,
twisting his signet ring.
He liked adding 's' to a word
and scoring more points
than the person who thought of it.

He wanted 'chinas'. He said they were ornamental
bricks from Derbyshire, hand-painted.
He cheated from principle, to open up new ground
for his family. Not 'God feeds the ravens' so much
as *Mundum mea patria est*. We were stuck

at the end of a lane in Sussex
for two winters. My father threw down
his high-scoring spelling mistakes and bluffs
and started counting.
He would have counted us all out
if we'd let him have the last word—

'aw' as in 'Aw, hell!', 'ex' with the 'x' falling
on the last Triple Letter Score.
We made him take everything back.
What was left in his hand counted against him.

Dégagé

Clothes were a kind of wit. You either
carried them off, or you looked ridiculous.
'Make a girl laugh,' said my father,
which I did. Whatever I put on
made me look even younger than my brother,
who was ten.

I tried every combination
of cravat and cardigan in my efforts to look
natural, *dégagé.* I dug my hand
into the pocket of my flannels
and felt the little rolls of pocket dust
under my fingernails—and remained a virgin.

My father's forty-seven suits
awaiting his pleasure in a separate dressing-room
were proof of his superior wit. Who else
had a white barathea dinner jacket
he never even wore, or turned-back cuffs
and no turn-ups on his trousers?

At fourteen I was nagging my grandmother
to make me shirts with fuller sleeves.
My jeans I wanted taken in and flared.
I was very keen on suede. 'You should be with someone
a full minute,' said my father,
'before you realize they're well dressed.'

I imagined it dawning on people
in sixty seconds flat
that I was his equal at last.
'Suppose you realize before that?' I asked,
wriggling my toes in my chisel-toed chukka boots.
'Probably queer,' said my father.

Snorkel

To my Brother

You carried the rattans and the towels.
I carried the windshield
and one of the old snorkels
with ping-pong balls for valves.

What happened to the other one
with yellow glass, the one that was dangerous?
We both wanted that one.
It didn't mist up. We slung ourselves

half way between heaven and earth
that summer—holding our breath
and diving for sand-dollars.
If we breathed out all the air in our lungs

we could grab another ten seconds
on the sea-bed. We spent half our lives
waiting for each other to come out of the water
so we could have our turn.

Three Quarters

I wasn't happy with aspects of my case.
I shut myself in the bathroom,
a three-sided looking-glass open like a book.
I couldn't understand my face. My nose stuck out.
I combed my hair down over my eyes
in search of a parting that would change all this.

I opened the mirror slowly, turning my head
from full to three-quarter face.
I wanted to stand three-quarters-on to the world,
near the vanishing point.
I sat in front of the sunray lamp
with pennies in my eyes. I dyed my skin

a streaky, yellowish brown with permanganate of potash.
I must have grown up slowly
in that looking-glass bathroom,
combing my hair straight down and pretending to wash.
I made myself dizzy raising my arms above my head
in a kind of surrender. No one else could get in.

No Particular Place To Go

O'Sullivan's Record Exchange
in the Peskett Street Market
was out of bounds to Lower Boys
on account of Miss O'Sullivan's taste
in music. We used to jive
in the listening booths
when she turned the volume up for us,
knowing we wouldn't buy.
It was the best she could do.
You couldn't hear that kind of thing
any other way in 1956. The overloaded wires
must have set fire to the partitioning.
They had to throw hundreds of
twisted 78's out on to the pavement.

O'Sullivan's Record Exchange,
its record-covered walls suspended
in their own flames, still seems to welcome me
with all my favourite tunes,
and Miss O'Sullivan
moving her arms over the turntables
like one who heals. When I'm caught
loitering in the new car park
off Peskett Street ten years from now
and taken for questioning, I'll know
what to expect: 'Look here old boy,
the past is out of bounds, you should know that.'
'But sir,' I'll say, 'where else is there to go
on these half-holidays?'

The Spring of Sheep

Pro-Plus Rapid Energy Tablets
gave me Extra Vitality
when I visited my girlfriend on her father's stud.
The double-backing local bus
took two hours to travel twenty miles.
When it passed our house
I nearly got off by mistake.
I noticed a roof I hadn't been on
and I wished I was up there with my gun.
My hands were shaking
as I thought of things to say:
how the enlargements had gone astray
and been pinned to the noticeboard,
how my tutor asked if it was Brigitte Bardot.
I practised laughing in the window of the bus,
but I laughed on the other side of my face
when I saw her riding her pony
in her Sloppy Joe.
We were sitting alone in the nursery,
waiting for her father's horse to appear on television.
My left hand felt numb,
but my right took leave of its senses
and set out for the unknown regions of her shoulders.
I watched through binoculars
as it lay there with altitude sickness.
If it was mine, how could I get it back in time
for dinner with her parents, bloodstock
and doping scandals? A gong
sounded somewhere in the house
and I leapt to my feet. Everyone was proud
of the gallant Citizen Roy
and my girlfriend ran over to the stables
to say goodnight. Head-over-heels with Pro-Plus,
I lay awake for hours, experiencing fierce
but tender feelings for the mattress
in a spare room hung with antique jigsaws:
'Les Generaux en herbe (The Future Generals)'
'Le Jeu de Balle (The Game of Balls)'
'Le Saut du Mouton (The Spring of Sheep)'.

A Start in Life

Of course I wanted to be an actor. I had the gold chain
like Alain Delon. I could lift one eyebrow.
I didn't wear any socks.
I came home from France
with a brush-cut and a sketch of myself
and my father said 'WHAT ARE YOU GOING TO DO?'

Work had this mad glint in its eye
which made me look away.
I practised my draw in the mirror.
'The honeymoon's over,' said my father.
'I don't care what you do
so long as it isn't a politician, a poof, or a tenor.'

I made a face, scanning the South Downs
for something easier.
On a good day I could see the Chanctonbury Ring
outlined against the horizon.
'I want to be an actor,' I said.
My father slapped his knee.

'No you don't,' he shouted. 'You don't give a damn
about the theatre, or me. You write poetry.
When I was your age I'd seen every play in London.
I wanted passionately to act.
Can you say that?' His widow's peak
was like a judge's black cap as he laid down the law.

'Acting's showing off,' I said to the Downs.
'It's the perfect cover for people like us
who can't do anything else.
It's better than nothing anyway.'
I walked in the garden, shaking one of his collars
till it fell to pieces in my hand.

I dried my eyes, but I never did land
the job he was looking for. I stayed where I was,
waiting for a last call to find me
putting on make-up in my dressing room—
'Five minutes please, Mr Williams'—
as if I could still go on

and make a start in life. I see the Downs even now
like a backdrop to the scene.
I put on different clothes and I see myself in action.
It feels like drawing a gun in slow motion
over and over again. I have the gold chain
like Alain Delon. I can lift one eyebrow.

Returning Soldier

He must be standing by a window, looking out
on a backdrop of Regent's Park. The sound
of carriage wheels on gravel, a woman's laugh.
As the lights come up, he moves to centre stage
to check his tie: the perfect kid-gloved cad.

In a government-issue busman's overcoat,
long in the sleeve,
a white arm-band for 'officer material',
he looks more like the wronged husband of the piece.
'Don't just do something, sit there'
was the word of command
to the men guarding Staines Railway Bridge
during the Phoney War.

As the dust settled I could see your father
stretched out beside the road, clutching a map of France.
I lifted him up
and propped him against the side of the jeep.
'Come on, sir', I said. 'Have a cigarette, sir.
You always have a cigarette when you wake up.'
He didn't get the picture at first.
He thought he had trodden in something cold
and fallen over backwards.
'I got my bastard left foot wet', he said.

Now that he's walking towards me in long shot,
limping a little from the war,
Now that he pauses for a moment in close-up,
lighting a cigarette,
I find myself playing the kid-gloved cad
to his returning soldier.
I'm sitting on the edge of my seat
to find out what I say.

Going Round Afterwards

His face was orange.
His widow's peak had been blacked in.
I knew it was him,
because he didn't speak.
'Congratulations!' I said.
'I didn't know you could cry.'
His dresser was holding
a pair of check trousers
underneath his chin. He let the legs
drop through a coathanger
and smiled at me deafly.
'It's just a trick', said my father.
'Anyone can do it.'
I stood there with my drink,
feeling the ingenious glamour
of being cramped, the mild delinquency
of things behind curtains—
shirts and cardigans
that should have been at home.
Did I have the guts?
And did you have to want it all that much
in order to go on?
His face came up from the wash basin
white and unwell again,
a trace of make-up underneath his ears.
His dresser was handing him
another pair of trousers,
holding them up off the floor
as my father stepped into them.

Death Of An Actor

i.m. Hugh Williams 1904–1969

1

Now that I am cold
Now that I look like him
I put on this warm grey suit of wool
In sympathy with my father.

Now that I'm alone
Now that I have come to this nice
Indifference
I sweep my hair straight back
The way he wore it during his life
And after he was dead
His fierce forehead
Still doubting the intelligence
Of those who approached where he lay.

2

Now that he is dead
Now that he is remembered
Unfavourably by some
For phrases too well cut
To fit their bonhomie
I wonder what he was like
This stiff theatrical man
With his air of sealed regret.
'I'd have made a first class tramp',
He told me once,
'If I'd had more money.'

Now that it is late
Now that it is too late
For filial piety
I can but thank him for
His bloody-mindedness.
Face expressionless with pain
He ordered me a suit in Savile Row

The very day he took
The last plunge backwards
Into secrecy and sweat.
'O Dad, can dead men swim?'

3
Gold on the doorstep, whose steps
Nag the sand-drifts.
Gold in the spittoon.

My father would sit on the steps
Emptying his shoe.
Pitchers of sand on each step.

If they went on, they would lead
Nowhere. Gold in a silver spoon.
My father's throat torn to sand.

4
Our first Christmas after the war
A triangular package
Arrived from his producer.

'Greetings from Emile Littler'
Said the message printed on the bar
Of a single coathanger.

5
Now that I have tucked myself in
To this deep basement calm
And the windows are sealed for winter,
Now that my life is organized
To absorb the shock
Of looking back at it,
I understand why he put such vast whiskies
Into the hands of his enemies
And I take back what I said.

Now that I am grown
Now I have children of my own
To offer me their own
Disappointed obedience
I feel for him.
Our children left us both
Because we sat so still
And were too wise for them
When they told us their best jokes.

6
My father was last to leave the stage
In *The Cherry Orchard* in 1966.
He said to his bookshelves,
'My friends, my dear good friends,
How can I be silent?
How can I refrain from expressing, as I leave,
The thoughts that overwhelm my being?'
His sister was calling him,
'The station . . . the train. Uncle,
Shouldn't we be going?'

7
The recording starts too late
To drown the sound of wheels. A little screen
Jerks upwards and the coffin
Wobbles towards us on rollers, like a diving board.
This is my father's curtain call. His white-ringed eyes
Flicker to the gallery as he bows to us. He bows
To his leading lady, then steps back again,
Rejoining hands with the cast.

In the dressing-room afterwards
He pours us all champagne:
'It's like a madhouse here. We're staffed by chumps.
The stage manager thinks the entire production
Stems from his control panel, like a cremation.
He's never heard of laughs. As for the set,
Tom says it's the old Jermyn Street Turkish Baths
Painted shit. Let's hope it doesn't run.'

8

Now that he is gone
Now that we have followed him this far
To a push-button crematorium
In unknown Golders Green
I think how near he seems, compared to formerly,
His head thrown back like that
Almost in laughter.
I used to watch him making up
In an underground dressing-room,
His head thrown back that way:
A cream and then a bright red spot
Rubbed down to a healthy tan.

Now that he is gone
Now that we have seen his coffin
Roll through those foul flaps
And a curtain ring down for the last time
On a sizeable man
I remember how calm he remained
Throughout the final scene,
Sitting bolt upright
On a windswept platform.
'The coldest place in the south of England',
He used to say—off on tour again
In one of his own plays.

9

Now that he has returned to that station
Where the leave-train is waiting
Blacked-out and freezing,
The smell of whisky lingers on my breath,
A patch of blue sky
Stings like a slap in the face.

Now that he isn't coming down
On the midnight train tonight, or any night,
I realize how far
Death takes men on from where they were
And yet how soon
It brings them back again.

10

Now that I'm the same age
As he was during the war,
Now that I hold him up like a mirror
To look over my shoulder,
I'm given to wondering
What manner of man it was
Who walked in on us that day
In his final uniform.
A soldier with two families?
An actor without a career?
'You didn't know who on earth I was,' he told me.
'You just cried and cried.'

Now that he has walked out again
Leaving me no wiser,
Now that I'm sitting here like an actor
Waiting to go on,
I wish I could see again
That rude, forgiving man from World War II
And hear him goading me.
Dawdling in peacetime,
Not having to fight in my lifetime, left alone
To write poetry on the dole and be happy,
I'm given to wondering
What manner of man I might be.

OXFORD POETS

Fleur Adcock
James Berry
Edward Kamau Brathwaite
Joseph Brodsky
Michael Donaghy
D. J. Enright
Roy Fisher
David Gascoyne
David Harsent
Anthony Hecht
Zbigniew Herbert
Thomas Kinsella
Brad Leithauser
Herbert Lomas
Derek Mahon
Medbh McGuckian
James Merrill
John Montague
Peter Porter
Craig Raine
Tom Rawling
Christopher Reid
Stephen Romer
Carole Satyamurti
Peter Scupham
Penelope Shuttle
Louis Simpson
Anne Stevenson
George Szirtes
Anthony Thwaite
Charles Tomlinson
Chris Wallace-Crabbe
Hugo Williams

also

Basil Bunting
W. H. Davies
Keith Douglas
Ivor Gurney
Edward Thomas